The Natural Histor[y]
Animal Clo[se-Ups]

Night
Animals

Barbara Taylor

PETER BEDRICK BOOKS

McGraw-Hill
Children's Publishing
*A Division of The **McGraw·Hill** Companies*

Published in the United States in 2003 by
Peter Bedrick Books, an imprint of
McGraw-Hill Children's Publishing,
A Division of The McGraw-Hill Companies
8787 Orion Place
Columbus, OH 43240

www.MHkids.com

ISBN 1-57768-964-X

Library of Congress Cataloging-in-Publication Data is on file with the publisher.

Text copyright © Barbara Taylor 2003
Photographs copyright © The Natural History Museum, London 2003
Photographs by Frank Greenaway

The moral rights of the author have been asserted

Database right Oxford University Press (maker)

All rights reserved. No part of this publication may be reproduced,
stored in a retrieval system, or transmitted, in any form or by any means,
without prior permission in writing of the publisher.

1 2 3 4 5 6 7 8 9 10 OXF 06 05 04 03 02

Printed in Hong Kong

Contents

We are night creatures.	6
Eyes, ears, and noses	8
Eagle owl	10
Bat	12
Hedgehog	14
Field mouse	16
Fox	18
Moth	19
Snail	20
Gecko	21
Glossary	22
Index	23

We are night creatures.

All sorts of animals come out at night. Some use the darkness to hide or to hunt. Others wait for the cool of night.

I am a barn owl. I am a nighttime hunter of small mammals, such as mice and voles.

We are banded snails. We dry out easily, so the damp night air helps us to survive.

I am a moth. I come looking for food and for a mate when it is dark.

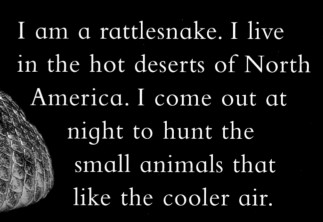

I am a rattlesnake. I live in the hot deserts of North America. I come out at night to hunt the small animals that like the cooler air.

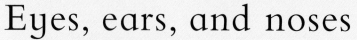

Eyes, ears, and noses

I am a pit viper. In total darkness I sense the hot bodies of my prey with the heat pits under my eyes.

I am a dormouse. My huge eyes, sensitive nose, and long whiskers help me to find my way around at night and escape danger.

I am a beetle. I use my long antennae to touch and smell things in the dark.

I am a red fox. My large ears work like funnels to catch as many sounds as possible in the quiet of the night.

I am a huge eagle owl.

I sleep during the day. At night I go hunting. I fly very quietly, so my prey cannot hear me coming.

My streaky brown feathers give me good camouflage.

I am the largest of all the owls. I am strong enough to attack animals as big as hares and mallard ducks.

I catch my prey with my needle-sharp talons.

My eyes are huge and round with big pupils, to let in lots of light. I have excellent hearing, too.

I am a squeaking bat.

I fly fast and low over water at night to catch insects. My wings fold neatly against my sides when I rest.

This "basket" of skin helps me to hold my prey.

My furry body keeps me warm in the cool of the night.

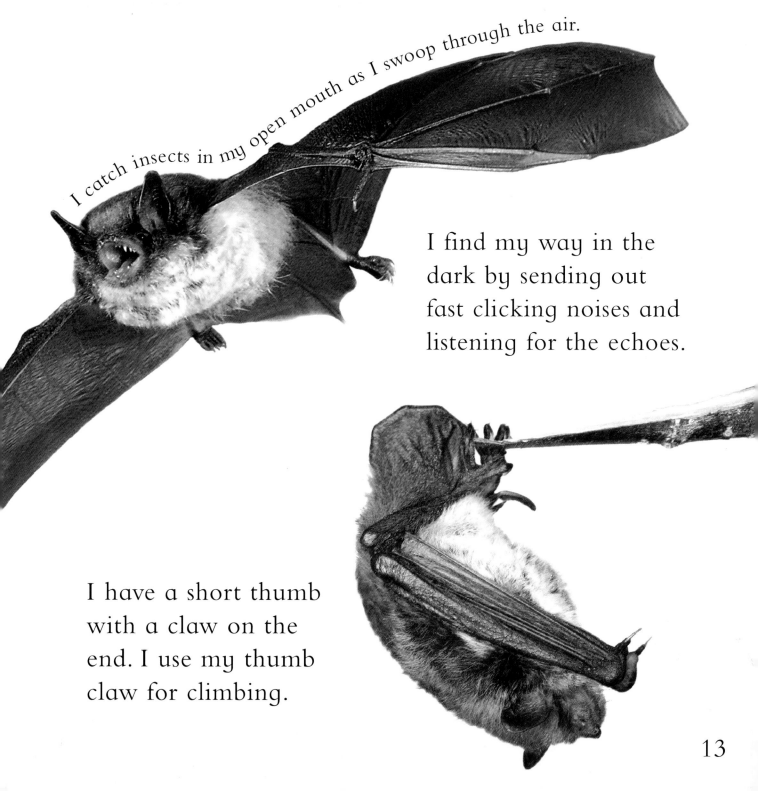

I catch insects in my open mouth as I swoop through the air.

I find my way in the dark by sending out fast clicking noises and listening for the echoes.

I have a short thumb with a claw on the end. I use my thumb claw for climbing.

I am a spiny hedgehog.

I move around at night when there are lots of beetles, grubs, worms, and slugs to eat. I sleep through the cold winter.

My sharp spines are hard, pointed hairs.

I can roll myself into a ball to protect my soft belly. I have about 7000 sharp spines!

I use my strong claws for climbing.

I am good at smelling and hearing things, but my vision is poor.

I am a shy field mouse.

I come out at night when it is harder for my enemies to see me. I can climb trees but I usually live on the ground.

My long, scaly tail helps me to balance.

I have large eyes and ears to help me to see and hear in the dark.

I am called a yellow-necked field mouse, because of the yellow fur across my throat.

I use my strong, gnawing teeth to eat seeds, insects, and fruit.

My long toes and claws help me to grip branches.

I am a cunning fox.

Although I hunt for prey, I may also find food in city garbage.

I use my four pointed teeth to kill my prey.

I can run fast using my tail for balance.

I am a fluttering atlas moth.

I am a male.
I flutter in the
night air looking
for a female.

My large wings are covered in tiny scales.

My feathery antennae pick up a female's scent.

I am a stretchy snail.

My slimy body dries out easily.
I generally come out at night,
when the air is cool and damp.

My slippery
slime helps me
to glide smoothly
over things.

I see with my two long tentacles.

I am a spotty gecko.

During the day I hide away in quiet, shady places. I come out at night to hunt for food.

My pupils close to a slit in bright light.

I grip the ground with my toes.

I wipe my big eyes clean with my tongue.

My ear holes are on the sides of my head.

I store fat in my tail. It helps me survive when food is hard to find.

Glossary

antennae Long, thin stalks on an insect's head, used for sensing things.

mammal An animal that feeds its young with mother's milk. Human beings are mammals.

prey An animal that is killed or eaten by another animal.

pupil The black hole in the middle of the eye, through which light enters.

talons The strong, thick claws of a bird of prey.

tentacle A long, flexible structure like an arm, near an animal's mouth. It may have suckers or stings.